I0503449

ISO 13485 – the Medical Device Quality Management System

CONTENTS

It is the responsibility of individuals, companies and organisations to implement the necessary legal and regulatory requirements relevant to their industry. Solo Validation Resources Limited will take no responsibility for the application or interpretation of this document.

ABOUT THIS BOOK

This book is broadly divided into 3 parts. The first part introduces the standard ISO 13485 and Quality management systems. Part two then examines the key are of Design controls and there application. Finally, a review of Quality Risk management is provided. In the first instance, providing safe and effective medical devices depends on sound basis' of design. However, how we see and rate risks also impacts the safety of products produced. A holistic approach to medical device manufacturing ensure Quality from design conception to commercial manufacturing. Following the principles within this short book will put the reader on the right tract.

1
INTRODUCTION

ISO 13485 is the quality management standard of choice for manufactures of medical devices.

Revised in 2016, ISO 13485:2016 "specifies requirements for a quality management system where an organisation needs to demonstrate its ability to provide medical devices and related services that consistently meet customer and applicable regulatory requirements."[1] The scope of the standard can apply to any organisation or company involved throughout the life-cycle of a product, including design and/or development, production, storage and distribution, installation, or servicing of a medical device and design and development or provision of technical or professional services. [1] International Standards Organisation, www.iso.org

The recent revision is designed to address recent developments in quality management and other updated regulations that relate to the industry. Improvements in the new version of the standard include broadening its applicability to include all organisations involved in the life cycle of the product, from the concept stage to end of life along with greater alignment with regulatory requirements and post-market surveillance including complaint handling.

ISO 13485:2016 is also used by suppliers or external vendors that provide QMS related management system-services.

Requirements of ISO 13485:2016 are applicable to organisations regardless of their size and regardless of their type except where explicitly stated. Wherever requirements are specified as applying to medical devices, the requirements apply equally to associated services as supplied by the organisation.

If any requirement in Clauses 6, 7 or 8 of ISO 13485:2016 is not applicable due to the activities undertaken by the organisation or the nature of the medical device for which the quality management system is applied, the organisation does not need to include such a requirement in its quality management system. For any clause that is determined to be not applicable, the organisation records the justification as part of their certification and quality management system.

The Process Approach

ISO 13485 is based on a process approach to quality management. A process is any activity that receives inputs and converts them to outputs.

For an organisation to function effectively, it has to identify and manage numerous linked processes. Furthermore, many processes impact other processes or downstream processes. The application of a system of processes within an organisation, together with the identification and interactions of these processes, and their management, can be referred to as the "process approach".

2

DIRECTIVES VERSUS STANDARDS

When it comes to regulated industries such as medical devices, it is first important to be familiar with some common terms and definitions and what they really mean. This chapter examines some key terms that are applied widely and relate to regulated industries. They include:

- DIRECTIVES

- STANDARDS

- NOTIFIED BODY
- COMPETENT AUTHORITY

Directives

Directives are legal requirements which must be met by manufacturers or other bodies within the industry. Directives are based on legislation and are issued at governmental level. It is important to note that standards such as ISO 13485 help companies meet the requirements set up in directives. (See harmonised standards below)

Standards

Standards are not always mandatory. However, they help manufacturers be compliant with directives/legislation. They also represent the current and best practice in the field of study/industry.

Harmonised standards are European standards prepared under a mandate from the European Commission, referenced in the official journal, and drafted so that compliance with their requirements relates to one or more essential requirements of the directive.

These standards have special status because, when a manufacturer can show that their products meet the requirements of the standard, there is a presumption that the product conforms to the essential requirements of the directive that is covered by the standard.

What is a Competent Authority?

When it comes to medical devices, a competent authority is the legally delegated authority mandated to monitor compliance to directives and legal requirements within the industry. The competent authority has the power to grant and revoke licenses.

Example of Competent Authorities:

• FDA (Food and Drug Administration) CFR Code of Federal Regulations – U.S.

• MHRA (Medicines and Healthcare Regulatory Agency - UK

• HPRA (Health Products Regulatory Agency) - Ireland

• JPAL (Japanese Regulations for Medical Devices) - Japan

What Is a Notified Body?

A notified body is a certification organisation which the national authority (the competent authority) of a member state designates to carry out one or more of the conformity assessment procedures described in the annexes of the medical devices directives.

The Medicines and Healthcare Products Regulatory Agency is the UK competent authority under the three directives. The Health Products Regulatory Authority is the Irish competent authority.

Organisations and Institutions

Many of the common acronyms that are referenced in literature relate to various standard setting organisations and industry representatives. Some of the more common bodies are listed below:

ISO: Internal Organisation for Standardisation

IMDR (F): International Medical Device Regulators Forum

ASTM: American Society for Testing and Materials

GHTF: Global Harmonisation Task Force

3

BASIC DEFINITIONS

(Source: Annex IX of Directive 93/42/EEC)

Intended Purpose

"Intended purpose" means the use for which the device is intended according to the data supplied by the manufacturer on the labelling, in the instructions and/or in promotional materials. (Chapter I section 1 of Annex IX of Directive 93/42/EEC)

Transient

Normally intended for continuous use for less than 60 minutes.

Short Term

Normally intended for continuous use for not more than 30 days.

Long Term

Normally intended for continuous use for more than 30 days.

Invasive Devices

A device which, in whole or in part, penetrates inside the body, either through a body orifice or through the surface of the body.

Body Orifice

Any natural opening in the body, as well as the external surface of the eyeball, or any permanent artificial opening, such as a stoma.

Surgically Invasive Device

An invasive device which penetrates inside the body through the surface of the body, with the aid of or in the context of a surgical operation.

Implantable Device

Any device which is intended:

- to be totally introduced into the human body or,
- to replace an epithelial surface or the surface of the eye,

by surgical intervention which is intended to remain in place after the procedure.

Any device intended to be partially introduced into the human body through surgical intervention and intended to remain in place after the procedure for at least 30 days is also considered an implantable device.

Medical Device

'Medical device' means any instrument, apparatus, appliance, material or other article, whether used alone or in combination, together with any accessories or software for its proper functioning, intended by the manufacturer to be used for human beings in the:

- diagnosis, prevention, monitoring, treatment or alleviation of disease or injury.

- investigation, replacement or modification of the anatomy or of a physiological process.

- control of conception which does not achieve its principal intended action by pharmacological, chemical, immunological or metabolic means.

A medical device may be assisted in its function by the following means:

'Active Medical Device' means any medical device relying for its functioning on a source of electrical energy or any source of power other than that directly generated by the human body or gravity.

'Active Implantable Medical Device' means any active medical device which is intended to be totally or partially introduced, surgically or medically, into the human body or by medical intervention into a natural orifice, and which is intended to remain after the procedure.

'Custom-Made Device' means any active implantable medical device specifically made in accordance with a medical specialist's written prescription which gives, under his responsibility, specific design characteristics and is intended to be used only for an individual named patient.

'Device Intended for Clinical Investigation' means any active implantable medical device intended for use by a specialist doctor when conducting investigations in an adequate human clinical environment.

'Intended Purpose' means the use for which the medical device is intended and for which it is suited according to the data supplied by the manufacturer in the instructions.

'Putting into Service' means making available to the medical profession for implantation.

Where an active implantable medical device is intended to administer a substance defined as a medicinal product within the meaning of Council Directive 65/65/EEC of 26 January 1965 on the approximation of provisions laid down by law, regulation or administrative action relating to proprietary medicinal products (6), as last amended by Directive 87/21/EEC (7), that substance shall be subject to the system of marketing authorisation provided for in that directive.

Where an active implantable medical device incorporates, as an integral part, a substance which, if used separately, may be considered to be a medicinal product within the meaning of Article 1 of Directive 65/65/EEC, that device must be evaluated and authorised in accordance with the provisions of this directive.

4

ISO 13485
&
REGULATIONS

In chapter 2 the special status of harmonised standards was described which allows companies meet the essential requirements of Directives. In Europe, EN ISO 13485:2013 helps companies meet the requirements of Directive 93/42/EEC on medical devices. This harmonised standard gives companies the "presumption of conformity" to complying with directives. EN ISO 13485 was published in February 2013 and harmonised in August 2013 to cover the three directives:

- 90/385/ECC– The Active Implantable Medical Devices Directive (AIMDI)

- 93/42/ECC – The Medical Devices Directive (MDD)

- 98/79/EEC – In Vitro Diagnostic MDD (IVDMDD)

In the United States, medical device manufacturers need to meet the requirements of 21 CFR Part 820 of FDA regulations. While ISO 13485 is not an actual requirement, many companies will seek certification to the standard to support the exporting of products.

In Australia, it is a regulatory requirement for manufacturers of medical devices to meet the requirements of ISO 13485.

In Canada, certification to ISO13485 is part of the regulatory requirements.

ISO 13485 is the quality management standard for medical devices and is the certification standard of choice for medical device manufacturers throughout the globe. It is a mandatory requirement for many countries and forms a central part of compliance in others. The content of ISO 13485 is interpretive (not prescriptive) which gives a degree of scope in how the requirements are applied and met within a company.

ISO 13485 provides both a sound and widely recognised basis in meeting regulatory compliance for medical devices. Based off ISO 19001 however, ISO 13485 is a standalone standard for medical devices.

ISO 9001 has requirements and themes relating to continual improvement and customer satisfaction. These have been modified for ISO 13485.

Main differences between ISO 9001 & ISO 13485

• Customer satisfaction is changed to customer feedback

• Extra requirements regarding procedures for ISO 13485

• Extra requirements for records ISO 13485 (e.g. retention)

• Continual improvement is restricted to continual improvement of the quality management system

ISO 13485 has extra requirements required for regulatory bodies such as post production review and management of advisory events.

ISO 13485 and ISO/TR 14969

ISO/TR 14969 is a technical report that is used for guidance on the application and implantation of ISO 13485. It is recommended for those responsible for the role out of ISO 13485 within their organisation. The content of ISO/TR 14969 is based on several established organisations such as the GHTF, ISO and input from regulatory bodies.

4

GETTING STARTED

What is a Quality Management System, QMS?

A Quality Management System is a framework by which quality and a clear set of objectives are realised and achieved. A QMS fosters improvement and facilitates implementation of change.

It is important to remember that a QMS is not a static activity. It should have a momentum and energy that makes it a living framework. Certification is a key benefit but if the QMS is truly lived by the company it benefits to go further.

Customer and Regulatory Focus

An understanding of the customer needs and requirements should be evident within the organisation and with the future vision of the company. The company should have an understanding of the regulatory landscape as this is subject to change over time. In turn the company should be positioned to respond to that change.

Leadership

To truly lead, one must be accepted in the hearts and minds of those they lead. Authentic leadership pays off. A leader should foster a sense of togetherness and common vision. A leader is anyone who influences or directs people either formally or informally. We are all leaders to some extent.

Involvement

Engagement by everyone across an organisation is now recognised as being key in the successful deployment of any Quality Management System. Everyone should have a voice within the company. The saying "we are only as strong as the weakest link" is very apt.

The Process Approach

ISO 9001 and ISO 13485 are standards that are based on process approaches. A process approach essentially utilises methods or standardised tools to help drive consistency. Some common tools include DFMEA /DMAIC/PDCA to name but a few.

Systems Management

This essentially means that systems are defined and described in writing along with the appropriate responses to expected issues that arise. Effective systems management must ensure that the various systems work in support of each other and communicate effectively with one another.

Decision Making

In order to make the right decision, the person empowered to make the decision must be informed. To be correctly informed one must have the correct details and facts available. In a manufacturing environment the facts are essentially data and the analysis of data.

During manufacturing or processing, data is generated as a result of monitoring and measurement of products and the related processes.

Supplier Management

Don't ruffle your suppliers' feathers. Security of supply is key in delivering products to customers or patients again and again, Raw materials or sub-components sourced from external suppliers must always be sourced at the right price and time with the emphasis on getting the best quality possible.

Continuous Improvement

For ISO 13485 continuous improvement refers to improving the effectiveness of the Quality Management System. It is harder to drive improvement of the product due to regulatory and practical requirements.

This is a key difference in contrast to ISO 19001:2008 as there is a requirement to continually improve both product and processes.

Eight Principles Of CGMP

1. Design and construct the facilities and equipment properly with due regard for regulations and international ISO standards.
2. Write and follow written procedures and instructions at all times.
3. Document work accurately and in a timely fashion.
4. Validate equipment systems and processes.
5. Monitor facilities and equipment.
6. Protect against contamination and maintain a clean environment.
7. Control raw materials, components and product related processes.
8. Conduct periodic audits.

The 8 principles of quality management are the same throughout:

- ISO 13485 medical devices

- ISO 9001 quality management

- ISO 18001 occupational health and safety

- ISO 14001 environmental management

Benefits of ISO 13485

- It is a QMS that helps meet the regulatory requirements.
- Quality used to be based on quality control eg. reactive, inspection etc.
- Quality assurance is preventative – process approaches, proactive.
- Enhanced reputation.
- More increasingly becoming an expected standard for medical devices.

6

STANDARD OVERVIEW

ISO 13485 has 8 Clauses or Sections which make up the structure of the standard.

Section 0 Normative References, Definitions and Terms

Section 1 Requirements of the Quality Management System (QMS)

Section 2 Normative References

Section 3 Terms and Definitions

Section 4 Requirements of the Quality Management System (QMS)

Section 5 Management Responsibility

Section 6 Resource Management

Section 7 Product Realisation

Section 8 Measurement, Analysis and Improvement

CLAUSE 1: SCOPE

This section refers to the scope and application of the standard.

- The organisation must be able to show its ability to provide medical devices to meet customer requirements and regulatory requirements
- A key aim of the standard is to allow harmonisation to regulatory requirements
- The scope of the QMS must relate to medical devices for a company to be able to use ISO 13485.

Some examples of what's in scope of the standard include (1) the manufacture of hip implants, (2) the design and manufacturing of in-vitro blood testing devices, (3) contact analytical testing (4) consultancy services.

The terms "where appropriate" and "if appropriate" are used throughout the standard, therefore, it should be met by the organisation unless a justification is documented.

CLAUSE 2: NORMATIVE REFERENCES

This clause states that when working with ISO 13485, refer to ISO 9000:2000 for fundamentals and vocabulary.

CLAUSE 3: TERMS AND DEFINITIONS

This clause provides terms and definitions. It is very useful in the early days of establishing and implementing ISO 13485 to ensure that terms and definitions are clearly understood.

CLAUSE 4: QUALITY MANAGEMENT SYSTEM

Clause 4 details the general requirements that relate to the quality management system, the documentation requirements and record requirements.

Clause 4 includes:

4.1 General requirements clause

4.2 Documentation requirements clause

CLAUSE 4.1 GENERA REQUIREMENTS

The organisation must implement a Quality Management System, or QMS in order to provide the framework and structure to achieve ISO 13485 roll-out and implementation. However, the role of the QMS does not stop there. After initial roll-out, the requirements of the standard must be maintained and determined to be effective on an on-going basis. The following processes should be documented:

- List of all processes
- Process interactions
- Monitoring of processes
- Resources to facilitate rollout of processes
- Measure and monitor effectiveness
- System of identifying improvements

CLAUSE 4.2: DOCUMENTATION REQUIREMENTS

When it comes to the regulated industries such as the medical device industry, every process and procedure must be documented. Documentation ensures that everyone is working in the same manner with the same procedures. However, documentation is more than just writing down procedures and processes. It is also concerned with how documents are controlled, how they are updated and how they are stored.

Electronic Document Management Systems

Electronic document management systems aka EDMS are now the norm and gold standard for most medium to large organisations. Many companies that provide medical device manufacturers with an EDMS can customise the system to match the business processes particular to an organisation. With configurable or customisable software, validation and proper verification is important to ensure the system operates as intended. There are also regulatory requirements that stipulate the expectations and requirements of such systems. For example, the application of electronic signatures and the presence of audit trials. FDA 21 CRF Part 11 details the requirements with regards to electronic records and electronic signatures. For medicinal products in Europe, GMP V4 Annex 11 specifies similar requirements.

Revision control is a key element of the Quality Management Systems in place in regulated industries. As the need for changes in the document arises, the controlled document can be amended/updated. With each update the version number revises also. Some companies will use alphabetic revision control and to a lesser extent numeric revision control (Version A, Version B or Version 01, Version 02).

Controlled documents should always have a version number or revision number electronically on each page of the document. This is similar to books which always list what edition they are. e.g. first edition or second edition.

Records

Records are generated through the application of processes and procedures. These records can be related in quality inspection and manufacturing. The integrity and quality of records relating to the manufacture of medical devices is important, as it plays a part in safe-guarding the patient or user. Records may also help in the investigation of any quality issues, complaints or adverse events that may arise.

Principles of Good Documentation Practices or GDP, should be applied to records. In particular, handwritten entries should always be accompanied by a signature and date. This is important as traceability must be maintained in the event of an issue or complaint.

CLAUSE 5: MANAGEMENT RESPONSIBILITY

Clause 4 includes:

5.1 Management Commitment

5.2 Customer Focus

5.3 Quality Policy

5.4 Planning

5.5 Responsibility, Authority and Communication

5.6 Management Review

5.1 Management Commitment

It is essential that top management have an authentic and tangible commitment to meeting regulations and the expectations of customers. Quality should be at the forefront of all of activities. Management should encourage discourse and communication on all matters relating to internal processes, quality and the QMS as a whole.

5.2 Customer Focus

Customer –patient/user/doctor/family member
Customer feedback is a requirement of ISO 13485 and as such the manufacturer must engage with the customer. In instances where a defective product is received, the manufacturer must have a complaints process to facilitate proper feedback, communication and investigation.

5.3 Quality Policy

Simple statement /1 pager or more

Often quality policies will be displayed in reception areas etc. Copies should be signed and revision controlled.

Quality policy must have a commitment to maintain the effectiveness of the QMS.

5.4 Planning

Top management must plan quality objectives and ensure they are implemented and effective.

Some examples of quality objectives include :

reduce rework by 10%

reduce scrap by 5%

have customer complaints reduced by 2% per year

5.5 Responsibility, Authority and Communication

Roles and responsibilities are defined.

Job descriptions are in place.

Organisational charts are in place and accurate.

5.6 Management Review

The purpose of management review is to ensure the effectiveness of the QMS. Inputs to management review include:

(a) Audit results

(b) Customer feedback

(c) Process performance and conformity

(d) Corrective and preventative actions

(e) Deviations

(f) Regulatory changes and revisions

CLAUSE 6 : RESOURCE MANAGEMENT

Clause 6 of ISO 13485 is concerned with human resources, infrastructure and work environment.

Clause 6 includes:

6.2 Human resources

6.3 Infrastructure

6.4 Work Environment

Human Resources

People are the key part of any QMS. Therefore, they should have the appropriate level of education, skill and experience. A culture of quality must be lived by everyone.

Training

People must be suitably trained. Training must be documented and consistent throughout an organisation. Training must be seen to be effective. Proper records of education and training must be kept.

Human intelligence, human creativity and human labour are all key inputs to any factory or company manufacturing medical devices. Therefore, an organisation must be properly resourced in order to function correctly, meet the regulatory requirements and customer expectations.

6.2 Human resources

"Change the people or change the people"

With any organisation, it is only as good as the people it has in its make-up. Therefore, the people, operators, engineers, managers etc. all contribute to the quality management system.

Clause 6.2.2 also specifies requirements with regards to competence, awareness and training.

The person should be matched to the job in terms of their qualifications, experience and training. Typically, job descriptions are used to drive and capture these requirements.

Nowadays, most multinational companies will ask for evidence of qualifications, training and experience. These documents are then held on file in the event of an audit. This is recommended practice for medical device companies. While the standard does not specifically call out the need to hold records of degrees and qualifications on file, the company or organisation needs to demonstrate the suitability of the person to their respective roles, and filing the qualification provides the easiest method.

6.3 Infrastructure

Infrastructure has the ability to impact the quality of products and services. Therefore, it must be fit for purpose. It is especially important if the organisation is involved with the manufacture of medical devices. The following element need to be considered with regards to infrastructure:

- Location of equipment and the operating environment
- Equipment installation and validation
- Utilities required for the operation of equipment and systems

- Layout of the factory – flow or raw materials, in-process materials and finished products
- Environmental systems such as HVAC and fire suppression systems

6.4 Work Environment

The work environment is also closely related to infrastructure within a given organisation and they can both affect or impact upon the quality of products manufactured. Risk to product quality and patients is minimised by understanding the work environment and how it can impact the product.

When the interactions and risks are understood, work can then be done to eliminate risks or at least control or monitor them.

Environmental conditions that can impact upon product quality include:

- Humidity
- Temperature

- Air quality
- Room pressure differentials (negative / positive)
- Air flow/velocity

CLAUSE 7 : PRODUCT REALISATION

Clause 7 includes:

7.1 Planning of product realisation

7.2 Customer-related processes

7.3 Design and development

7.4 Purchasing

7.5 Production and service provision

7.6 Control of measuring devices

7.1 Planning of Product Realisation

Product realisation can be defined as a collection of processes and body of work that delivers a product or service to the customer. Remember, when it comes to medical devices, customers can be patients or users such as doctors and nurses.

It should be noted that organisations can opt to exclude specific requirements, in cases where product realisation is not applicable. However, any such exclusion should be based on sound rationale with the case clearly documented. An example of this may be where design and development is not conducted by the manufacturer e.g. contract manufacturers.

Planning is an often underestimated but remains a key element of product realisation. If adequate time and resources are given to planning, it makes all other processes run smoother, and therefore should help to produce improved products and services.

7.2 Customer-Related Processes

There are 3 elements that feed into customer-related processes. They include the following:

Determining the requirements related to the product Clause 7.2.1

Review of requirements relating to the product- Clause 7.2.2

Customer communication-clause 7.2.3

Customer requirements are typically captured in a User Requirements Specification. A requirements specification (URS) documents all of the desired attributes of a product or service. They can be made up by a combination of CQAs, regulatory requirements and design requirements. A URS can then form the basis for review of the product or service requirements.

With regard to customer communication, it is important to remind ourselves that we are concerned with ISO 13485 which as we very well know by now is the standard for medical devices. Therefore, having the right information available to the customer, patient or end user is important. When additional information needs to be transmitted or updates to information need to be communicated, an advisory note can be issued. Another important aspect of customer communication is customer feedback. This communication can be made up of positive feedback from the customer or users, or when there is a query with regard to a product or service. Therefore, processes or systems must be in place to make communication between customer and company both effective and timely.

7.3 Design and Development

Design and Development Verification and Validation ensure that the product is designed, developed and subsequently manufactured meeting all the customer requirements, regulatory requirements and business requirements. These requirements are classed as inputs to the design and development, and verification and validation ensure the inputs have been adequately taken into account.

The design and development testing sometimes replicate the commercial applications of the medical device, hence providing a realistic challenge in order to have confidence in the medical device.

Design Control

Design control is a necessary practice that ensures good engineering principles are maintained throughout the design phase of a product. It also refers to the continual design and development of the product through its very lifecycle. The design and development files and history must be controlled and maintained, with any changes properly assessed, tested and documented.

7.4 Purchasing

Bearing in mind that a quality management system considers all aspects of an organisation's functioning, purchasing and procurement of materials necessitates putting robust controls in place. Simply put, a purchasing process must exist.

7.5 Production and Service Provision

This requirement of ISO 13485 is an extensive section with a great deal of importance associated with it. As we are dealing with the manufacture of medical devices (or other activity associated with medical devices) there are specific requirements for sterile products. If a product is sterile, its use or application is likely to be associated with greater risks to the patient. Therefore, extra safeguards must be in place for sterile medical devices. Key sections of Clause 7.5 include:

(1) control of production and service provision – both general and specific requirements, (2) specific requirements for sterile medical devices, (3) validation of equipment and processes for production and service provision, (4) traceability and identification, (5) preservation of product controls with regard to monitoring and measuring medical devices.

7.6 Control of Measuring Devices

This clause requires an organisation to identify what monitoring and measuring is required and to ensure the product or service meets the customer requirements. A calibration procedure must also be maintained to ensure the equipment is accurate and reliable.

Calibration must ensure that:

- Equipment used to verify product quality is calibrated to a periodic schedule.

- The calibration is performed to an international standard.

- The calibration status of the equipment is recorded and visible.

- The equipment must be located within a suitable area in order to maintain accurate and reliable results.

If an organisation uses any computer software to monitor or measure outputs, the software must be verified before use via the appropriate validation and qualification activities.

CLAUSE 8 : MEASUREMENT ANALYSIS

Clause 8 includes:

8.1 General requirements

8.2 Monitoring and measurement

8.3 Control of nonconforming products

8.4 Analysis of data

8.5 Improvement

8.1 General Requirements

Measurement, analysis and improvement are the key themes of clause 8. As with all medical devices, inspection and testing both during manufacturing and post manufacturing is necessary to ensure products and services function as intended and without defects. With any type of measurement or inspection analysis, the method used to complete the testing is critical. The method must be fit for purpose, and the equipment must be suitable. This "method validation" typically is done during the design and development phase.

8.2 Monitoring and Measurement

Monitoring and measurement are dependant on the information or feedback provided from various sources. The most important feedback is the post-production feedback that is gathered from customers or the end user. Again, this occurs over the whole lifetime of the product or service in question. There are a number of methods that can be used to obtain feedback. Some examples include:

-Customer surveys

-Customer complaints

-Review of regulatory databases such as MAUDE.

-Repair and servicing information

8.3 Control of Nonconforming Product

Non-conforming product presents a risk to patients or users of medical devices. When a situation arises where non-conforming product is manufactured or detected through inspection processes, the product must be controlled and segregated to prevent unintended use or distribution.

Some examples resulting in non-conformance are:

When a manufacturing process drifts outside its validation window or operating parameters.

A certificate of analysis for a raw material is not provided by the supplier or the results are out of specification.

In-process testing was not completed at the defined intervals.

Training of personnel completing tests is not current or is inadequate.

8.4 Analysis of Data

In any engineering activity, data and the quality of the data is a key factor in making the right decisions. Provided the data collected is relevant and accurate, analysis of data can provide important insights into process performance, quality control and product functionality.

Data should be collated in a consistent way and controlled by a procedure. When it comes to medical device manufacturing, the sources and types of data are multiple. Data can be generated from in-process testing and data can be generated from end of line testing aka finished product testing.

8.5 Improvement

ISO 13485 fosters a culture of continual improvement. As we have seen, each activity can be described as a process. For example, a manufacturing process, a procurement process, a complaints process. The set of processes that make up the quality management system need to be continually reviewed to ensure they are suitable and effective for the task at hand. Typical tools used to keep improvement in mind include:

- Review of the quality policy and quality objectives

- Frequency and category of corrective and preventative actions (CAPA's)

- Customer complaints

- Management review input

7

CE MARKING

In Europe a QMS is required for CE marking of a medical device that is placed on the market in the EU.

ISO 13485:2003 is a harmonised standard that can be used by companies to show conformity of their QMS to requirements of directives. EN ISO 13485:2012 was harmonised in August 2012. This allows compliant companies receive an EC Declaration of Conformity.

Summary of the CE Requirements

- Manufacturers of class I devices or their authorised representatives must:
  - review the classification rules to confirm that their products fall within class I (Annex IX of the Directive)
  - check that their products meet the essential requirements (Annex I of the Directive)

- notify the competent authority, in advance, of any proposals to carry out a clinical investigation to demonstrate safety and performance of a device as required by the regulations

- obtain notified body approval for sterility or metrology aspects of their devices and where applicable prepare relevant technical documentation

- Draw up the 'EC Declaration of Conformity' (below) before applying the CE marking to their devices

- Register with the competent authority

- Implement and maintain corrective action and vigilance procedures including a systematic procedure to review experience gained in the post-production phase

- Make available relevant documentation on request for inspection by the competent authority.

In Europe, all medical devices must bear the CE marking of conformity (see Annex XII) of the directive) when they are placed on the market and/or put into service.

The CE marking must appear in a visible, legible and indelible form on the device or its sterile pack, where practicable and appropriate, and where applicable on any instructions for use and sales packaging.

For 'sterile' and 'measuring' devices, the CE marking must be accompanied by the identification number of the notified body that has acted under the relevant conformity assessment procedure.

EC Declaration of Conformity

In order to affix the CE marking, the manufacturer or their authorised representative must follow the EC declaration of conformity procedure referred to in Annex VII of the directive.

This procedure must be completed prior to placing the device on the market.

The 'EC declaration of conformity' is the procedure whereby the manufacturer or their authorised representative prepares the required technical documentation, puts into place corrective action and vigilance procedures and declares that the products meet the requirements set out in the directive.

Technical Documentation

The technical documentation should be prepared following review of the essential requirements and must cover all of the following aspects:

Description. A general description of the product, including any variants (for example names, model numbers and sizes).

Raw Materials and Component Documentation: Specifications including, as applicable, details of raw materials, drawings of components and/or master patterns and any quality control procedures.

Intermediate Product and Sub-Assembly Documentation: Specifications including appropriate drawings and/or master patterns, circuits, and formulation specification; relevant manufacturing methods and any quality control procedures.

Packaging and Labelling Documentation: Packaging specifications and copies of all labels and any instructions for use.

Design Verification: The results of qualification tests and design calculations relevant to the intended use of the product, including connections to other devices in order for it to operate as intended.

Risk Analysis: The results of risk analysis to review whether any risks associated with the use of the product are compatible with a high level of protection of health and safety and are acceptable when weighed against the benefits to the patient or user. If biocompatibility is relevant – for example for skin contact and invasive devices – a compilation and review of existing data or test reports based on the relevant standards is required.

Compliance with the Essential Requirements and Harmonised Standards: A list of relevant harmonised standards (for example sterilisation, labelling and information, biocompatibility, electrical safety, risk analysis, product group standards) which have been applied in full or in part of the products. If relevant harmonised standards have not been applied in full, then additional data will be required, detailing the solutions adopted to meet the relevant essential requirements of the directive.

The manufacturer may choose to prove conformity with the essential requirements of the directive through the use of their own standards and/or other relevant published standards (ISO, EN, BS). However, the use of such standards does not give similar, immediate presumption of conformity to the essential requirements of the directive. Therefore, using a harmonised standard provides greater protection to the manufacturer.

8

DEVICE CLASSIFICATIONS

The manufacturer, in preparing for CE marking, should first determine if their product falls within the scope of the directive or national regulation, either as a medical device or as an accessory to a medical device, as defined in Article 1 of directive 93/42/EEC and Article 2 of the regulation.

In order to be classified as a medical device, the product should have a medical purpose and its primary mode of action will typically be physical.

Level of Risk

General medical devices and related accessories must be classified into one of four classes, which are based on the perceived risk of the device to the patient or user. The classification of a device determines the conformity assessment options that are applicable to the device, with higher risk devices undergoing higher levels of assessment.

Class	Risk level
I	Low Risk
IIa	Medium Risk
IIb	Higher Risk
III	Highest Risk

Table: Device Classification and Risk Level

Classification Rules

There are eighteen rules outlined in Annex IX of the directive and related regulation that lay down the basic principles of classification.

In MEDDEV 2.4/1 Rev. 8, these rules are further explained and descriptive examples are provided. The eighteen rules are subdivided into four groups as follows:

Rules	Device Type
Rules 1 – 4	Non-invasive Devices

Rules 5 – 8	Invasive Devices
Rules 9 – 12	Active Devices
Rules 13-18	Special Rule e.g. devices containing tissue of animal origin, drug-device combinations

Table: Rules Corresponding to Device Type.

Annex IX and related guidance documents outline a number of key characteristics, listed below, that must be considered to correctly classify a device using the eighteen classification rules:

General Principles of Device Classification

- Medical devices are defined as articles which are intended to be used for a medical purpose. It is the intended purpose that determines the class of device and not the particular technical characteristics of the device. The intended purpose of the device should be substantiated (if required) and be representative of the technical characteristics of the device.

- It is the intended and not the accidental use of the device that determines its class.

- It is the intended purpose assigned by the manufacturer to the device that determines the class of device and not the class assigned to other similar products.

- Accessories are classified separately from their parent device.

- The mode of action of a medical device should be clear and evidenced with appropriate data to confirm this mode of action.

- If the device can be classified according to several rules then the highest possible class applies.

- Multipurpose equipment which may be used in combination with medical devices are not themselves classed as medical devices unless the manufacturer places them on the market with the specific intended purpose as a medical device.

- If the device is not intended to be used solely or principally in a specific part of the body, it must be considered and classified on the basis of the most critical specified use.

9

SUMMARY OF RULES

(Source: Guidelines Relating To The Application Of
The Council Directive 93/42/EEC On Medical Devices, MEDDEC
2.4/Rev.9 June 2010)

Rule 1

Rule 1: All non-invasive devices are in Class I, unless one of the other 17 rules apply. This is a fallback rule applying to all devices that are not covered by a more specific rule.

This is a rule that applies in general to devices that come into contact only with intact skin or that do not touch the patient.

Some non-invasive devices are indirectly in contact with the body and can influence internal physiological processes by storing, channeling or treating blood, other body liquids or liquids which are returned or infused into the body or by generating energy that is delivered to the body. These must be excluded from the application of this rule and be handled by another rule because of the hazards inherent in such indirect influence on the body.

Rule 2

Rule 2: All non-invasive devices are in Class I, unless one of the other 17 rules apply.

These types of devices must be considered separately from the non-contact devices of Rule 1 because they may be indirectly invasive. They channel or store substances that will eventually be administered to the body. Typically these devices are used in transfusion, infusion, extracorporeal circulation and delivery of anaesthetic gases and oxygen.

In some cases devices covered under this rule are very simple gravity activated delivery devices.

Rule 2: *All non-invasive devices intended for channelling or storing blood, body liquids or tissues, liquids or gases for the purpose of eventual infusion, administration or introduction into the body are in Class IIa:*

- if they may be connected to an active medical device in Class IIa or a higher class,

-if they are intended for use for storing or channelling blood or other body liquids or for storing organs, parts of organs or body tissues.

- in all other cases they are in Class I.

Rule 3

Rule 3: Non-invasive devices that modify biological or chemical composition of blood, body liquids or other liquids intended for infusion into the body.

These types of devices must be considered separately from the non-contact devices of Rule 1 because they are indirectly invasive. They modify substances that will eventually be infused into the body. This rule covers mostly the more sophisticated elements of extracorporeal circulation sets, dialysis systems and autotransfusion systems as well as devices for extracorporeal treatment of body fluids which may or may not be immediately reintroduced into the body, including, where the patient is not in a closed loop with the device.

Rule 3: *All non-invasive devices intended for modifying the biological or chemical composition of blood, other body liquids or other liquids intended for infusion into the body are in Class IIb,* *unless the treatment consists of filtration, centrifugation or exchange of gas or heat, in which case they are in Class IIa.*

These devices (Rule 3) are normally used in conjunction with an active medical device covered under Rule 9 or Rule 11.

Filtration and centrifugation should be understood in the context of this rule as exclusively mechanical methods.

Rule 4

Rule 4: Non-invasive devices which come into contact with injured skin.

This rule is intended to primarily cover wound dressings independently of the depth of the wound. The traditional types of products, such as those used as a mechanical barrier, are well understood and do not result in any great hazard. There have also been rapid technological developments in this area, with the emergence of new types of wound dressings for which non-traditional claims are made, e.g. management of the micro-environment of a wound to enhance its natural healing mechanism.

More ambitious claims relate to the mechanism of healing by secondary intent, such as influencing the underlying mechanisms of granulation or epithelial formation or preventing contraction of the wound. Some devices used on breached dermis may even have a life-sustaining or lifesaving purpose, e.g. when there is full thickness destruction of the skin over a large area and/or systemic effect.

Dressings containing medicinal products which act ancillary to the dressing fall within Class III under Rule 13.

Rule 4: *All non-invasive devices which come into contact with injured skin:*
- are in Class I if they are intended to be used as a mechanical barrier, for compression or for absorption of exudates,
- are in Class IIb if they are intended to be used principally with wounds which have breached the dermis and can only heal by secondary intent.

Products covered under this rule are extremely claim sensitive, e.g. a polymeric film dressing would be in Class IIa if the intended use is to manage the micro-environment of the wound or in Class I if its intended use is limited to retaining an invasive cannula at the wound site. Consequently it is impossible to say a priori that a particular type of dressing is in a given class without knowing its intended use as defined by the manufacturer. However, a claim that the device is interactive or active with respect to the wound healing process usually implies that the device is in Class IIb.

Most dressings that are intended for a use that is in Class IIa or IIb, also perform functions that are in Class I, e.g. that of a mechanical barrier. Such devices are nevertheless classed according to the intended use in the higher class.

For such devices incorporating a medicinal product or a human blood derivative see Rule 13 or animal tissues or derivatives rendered non-viable see Rule 17.

Rule 5

Rule 5: Devices invasive with respect to body orifices.

Invasiveness with respect to the body orifices (ear, mouth, nose, eye, anus, urethra and vagina) must be considered separately from invasiveness that penetrates through a cut in the body surfaces (surgical invasiveness). For short term use, a further distinction must be made between invasiveness with respect to the less vulnerable anterior parts of the ear, mouth and nose and the other anatomical sites that can be accessed through natural body orifices.

Surgically created stoma, which for example allows the evacuation of urine or faeces, should also be considered as a body orifice.

Devices covered by this rule tend to be diagnostic and therapeutic instruments used in particular specialities (ENT, ophthalmology, dentistry, proctology, urology and gynaecology).

Rule 5: *All invasive devices with respect to body orifices, other than surgically invasive devices and which are not intended for connection to an active medical device or which are intended for connection to an active medical device in Class I:*

- are in Class I if they are intended for transient use,

- are in Class IIa if they are intended for short term use except if they are used in the oral cavity as far as the pharynx, in an ear canal up to the ear drum or in a nasal cavity , in which case they are in Class I,

- are in Class IIb if they are intended for long term use,

except if they are used in the oral cavity as far as the pharynx, in an ear canal up to the ear drum or in a nasal cavity and are not liable to be absorbed by the mucous membrane, in which case they are in Class IIa.

All invasive devices with respect to body orifices, other than surgically invasive devices, intended for connection to an active medical device in Class IIa or a higher class, are in Class IIa.

Rule 6

Rule 6: Surgically invasive devices intended for transient use (< 60 minutes)

This rule primarily covers three major groups of devices: devices that are used to create a conduit through the skin (needles, cannulae, etc.), surgical instruments (scalpels, saws, etc.) and various types of catheters, suckers, etc.

This rule primarily covers three major groups of devices: devices that are used to create a conduit through the skin (needles, cannulae, etc.), surgical instruments (scalpels, saws, etc.) and various types of catheters, suckers, etc.

Rule 6: *All surgically invasive devices intended for transient use are in Class IIa unless they are:*

-intended specifically to control, diagnose, monitor or correct a defect of the heart or of the central circulatory system through direct contact with these parts of the body, in which case they are in Class III

-reusable surgical instruments, in which case they are in Class I

-intended specifically for use in direct contact with the central nervous system, in which case they are in Class III,

- intended to supply energy in the form of ionising radiation in which case they are in Class IIb,

- intended to have a biological effect or to be wholly or mainly absorbed in which case they are in Class IIb,

- intended to administer medicines by means of a delivery system, if this is done in a manner that is potentially hazardous taking account of the mode of application, in which case they are Class IIb.

Rule 7

Rule 7: Surgically invasive devices intended for short-term use (>60 minutes, <30 days).

These are mostly devices used in the context of surgery or post-operative care (e.g. clamps, drains), infusion devices (cannulae, needles) and catheters of various types.

Rule 7: *All surgically invasive devices intended for short term use are in Class IIa unless they are intended:*
- either specifically to control, diagnose, monitor or correct a defect of the heart or of the central circulatory system through direct contact with these parts of the body, in which case they are in Class III,
- or specifically for use in direct contact with the central nervous system, in which case they are in Class III,
- or to supply energy in the form of ionising radiation in which case they are in Class IIb,
- intended to have a biological effect or to be wholly or mainly absorbed in which case they are in Class III, - or to undergo chemical change in the body, except if the devices are placed in the teeth, or to administer medicines, in which case they are Class IIb.

Rule 8

Rule 8: Implantable devices and long-term surgically invasive devices (> 30 days). These are mostly implants in the orthopaedic, dental, ophthalmic and cardiovascular fields as well as soft tissue implants such as implants used in plastic surgery.

Rule 8: *All implantable devices and long-term surgically invasive devices are in Class IIb unless they are intended:*
- to be placed in the teeth, in which case they are in Class IIa,
- to be used in direct contact with the heart, the central circulatory system or the central nervous system, in which case they are Class III,
- to have a biological effect or to be wholly or mainly absorbed, in which case they are in Class III,
- or to undergo chemical change in the body, except if the devices are placed in the teeth, or to administer medicines, in which case they are in Class III.
- Directive 2003/12/EC introduced a derogation from this rule, reclassifying breast implants in Class III
Directive 2005/50/EC introduced a derogation from this rule, reclassifying hip, knee and shoulder joint replacements in Class III

Rule 9

Rule 9: Active therapeutic devices intended to administer or exchange energy.

Devices classified by this rule are mostly electrical equipment used in surgery such as lasers and surgical generators. In addition there are devices for specialised treatment such as radiation treatment. Another category consists of stimulation devices, although not all of them can be considered as delivering dangerous levels of energy considering the tissue involved.

Rule 9: *All active therapeutic devices intended to administer or exchange energy are in Class IIa*

unless their characteristics are such that they may administer or exchange energy to and from the human body in a potentially hazardous way, taking account of the nature, the density and the site of application of the energy, in which case they are in Class IIb. All active devices intended to control or monitor the performance of active therapeutic devices in Class IIb or intended to influence directly the performance of such devices are in Class IIb.

Rule 10

Rule 10: Active devices for diagnosis. This primarily covers a whole range of widely used equipment in various fields, e.g. ultrasound diagnosis, capture of physiological signals and therapeutic and diagnostic radiology.

Rule 10: *Active devices intended for diagnosis are in Class IIa:*
- if they are intended to supply energy which will be absorbed by the human body, except for devices used to illuminate the patient's body, in the visible spectrum,
- if they are intended to image in vivo distribution of radiopharmaceuticals,
- if they are intended to allow direct diagnosis or monitoring of vital physiological processes,
unless they are specifically intended for monitoring of vital physiological parameters, where the nature of variations is such that it could result in immediate danger to the patient, for instance variations in cardiac performance, respiration, activity of CNS in which case they are in Class IIb.

Active devices intended to emit ionising radiation and intended for diagnostic and therapeutic interventional radiology including devices which control or monitor such devices, or which directly influence their performance, are in Class IIb.

Rule 11

Rule 11: Active devices intended to administer and/or remove medicines, body liquids or other substances to or from the body. This rule is intended to primarily cover drug delivery systems and anaesthesia equipment.

Rule 11: *All active devices intended to administer and/or remove medicines, body liquids or other substances to or from the body are in Class IIa, unless this is done in a manner:*
- that is potentially hazardous, taking account of the nature of the substances involved, of the part of the body concerned and of the mode of application, in which case they are in Class IIb.

Rule 12

Rule 12: All other active devices. This is a fallback rule to cover all active devices not covered by the previous rules.

Rule 12: *All other active devices are in Class I*

Special Rules 12-18

Rule 13: Devices incorporating, as an integral part, a medicinal product or a human blood derivative (See MEDDEV. 2.1/3 for further guidance).

Rule 14: Devices used for contraception or prevention of sexually transmitted diseases.

Rule 15: Specific disinfecting, cleaning and rinsing devices.

Rule 16: Devices to record X-ray diagnostic images.

Rule 17: Devices utilising animal tissues or derivatives.

Rule 18: Blood bags.

10

DESIGN CONTROLS

Medical devices come in all shapes and sizes with different levels of complexity and risk. They range from simple devices such as bandages, plasters and urine test strips to automated diagnostic devices, orthopaedic implants, bone screws and artificial organs. Manufacturing companies vary in size, structure, and in their approach to design, development and management practices. All of these elements influence how design controls are implemented and how effective they are. However, an understanding of the user needs, patient needs and the design control requirements is essential to all manufacturers. It leads to better project outcomes and helps foster better communication and awareness delivering a quality and product that is fit for purpose.

Design controls are a collection of practices and procedures that are incorporated into the design and development process for a product such as a medical device. It provides a structure and clear path from user needs assessment to product delivery through a step-by-step process. Design controls ensure proper assessment of the design is completed during the design and development phase. It highlights technical issues, conflicts or deficiencies in design input requirements and allows them to be addressed early on in the process. Fixing a design issue early on reduces the cost of doing so at a later point and ensures the resultant design is appropriate for its intended use. Bringing a formal review process (design control) to the table assists engineers and managers in engaging with decisions and understanding them better. It also ensures that when future changes are made, they are documented and reviewed adequately with proper consideration to the design inputs.

Design controls are a requirement of quality systems such as 21 CFR Part 820 (medical devices), and for certain classes of devices and per ISO 13485 - Quality Management Systems.

Benefits of a Design Control System:

- The intended use of the device is documented and approved
- It ensures inputs align with outputs
- It creates a design "standard" and a "process" to allow benchmarking and consistency within an organisation

Design Controls and ISO 13485 Quality Management System

Clause 7 of ISO 13485 specifies the requirements for design and development of devices as part of the product realisation process. It should be noted that organisations can opt to exclude specific requirements of ISO 13485, in cases where product realisation is not applicable. However, any such exclusion should be based on sound rationale with the technical case clearly documented. An example of this may be where design and development are not conducted by the manufacturer e.g. contract manufacturers. Clause 7 (product realisation) of ISO 13485 details requirements for design and development controls. Clause 7 includes the following subparts:

Clause 7.1 Planning of product realisation

Clause 7.2 Customer-related processes

Clause 7.3 Design and development

Clause 7.4 Purchasing

Clause 7.5 Production and service provision

Clause 7.6 Control of measuring devices

Section 7.3 (Design and development) comprises:

Clause 7.3.1 Design and Development Planning

Clause 7.3.2 Design and Development Inputs

Clause 7.3.3 Design and Development Outputs

Clause 7.3.4 Design and Development Review

Clause 7.3.5 Design and Development Verification

Clause 7.3.6 Design and Development Validation

Clause 7.3.7 Control of Design and Development Changes

Definitions

Change Management: a management process where changes to the product, process, facilities or utilities are assessed, planned and reviewed as part of a formal systematic process.

Corrective and Preventative Action (CAPA): when an unplanned or adverse event happens, a corrective and preventative action can be implemented.

Design Phase Review: a process of evaluating the design requirements against the ability of it to deliver the intended device.

Design History File (DHF): an approved list of records that describe the design history of a medical device.

Design Input: the physical and performance requirements of a device that are the basis for the device design.

Design Output: the results of a design effort at each design phase and at the end of the total design effort. The finished design output is the basis for the device master record. The total finished design output consists of the device, its packaging and labelling, and the device master record.

Design Verification: confirmation by examination and provision of objective evidence that specified requirements have been fulfilled.

Design Validation: establishing by objective evidence that device or product specifications conform to user needs and intended use(s) defined in design documentation.

Device Master Record (DMR): a compilation of records containing the procedures and specification for a device. The contents of a DMR can contain local procedures such as SOPs and work instructions along with global or divisional specifications used to detail manufacturing processes, intermediate product or final product.

Design Phase Review: a documented, comprehensive, systematic examination of a design to evaluate the adequacy of the design requirements, the capability of the design to meet those requirements and to identify problems.

Specification: specification means any requirement to which a product, process, service, or other activity must conform.

Validation: validation means confirmation by examination and provision of objective, documented evidence that the particular requirements for a specific intended use can be consistently fulfilled.

Application of Design Controls

Design controls can be applied to any product development process. When the design input has been reviewed and the design input requirements are determined to be acceptable, the process of creating the device design begins. Product specifications are drafted and compared to the design input requirements. They then become the input for the next step in the design process. In the development and drafting of product specifications (e.g. critical quality attributes etc.) due regard must be given to product standards and industry best practices such as ISO and ASTM bodies. For example a catheter manufacturer should develop products with reference to ISO 10555 - intravascular catheters - sterile and single.

The Phase Approach to Design Control

The term "phase approach" is often used when describing the design control process. It simply means that a sequence of tasks needs to be completed, reviewed and approved during the development cycle of a product or medical device. Tasks are grouped into phases or stages. At the beginning, technical issues relating to design input requirements may need to be addressed with solutions identified. Often a range of solutions can be available, utilising different technologies. These different solutions then go on to be reviewed at the design selection process. At design selection, the project team must choose and justify a particular solution. The next phase (such as design verification and validation) ensures that the design is transferred to product launch and commercial supply - no oversights or deviations in the design intent occur. It also ensures that the device meets the user needs and intended uses (design inputs).

Managing Change

Changes made during design control are managed via document control procedures. For products built for commercial sale, the change management process is used to document and manage changes to the validated state of the process or the design of the product itself. While there may be more "flexibility" to make changes during the design phase of a project, diligence must be applied to any proposed change. Changes should be assessed by a multidisciplinary team with a management review.

Risk Management

Risk management involves the systematic application of management policies, practices and procedures that identify, analyse, control and monitor risk.

It is important to recognise that risk management should begin at the outset of the design and development phase of a project. The first step is to identify the user needs and intended use and application of the device. At the design input phase and design selection phase, risk assessments should be in a mature state. This allows the review of potential risks relating to the design of the product. Unacceptable risks can be dealt with by means of revisiting the design or introducing controls or mitigations in order to reduce the risks to acceptable levels. Following on from the design and development phase, the design verification, validation and transfer phases, or the clinical readiness phase, risk management activities and acceptability of the residual risk become the focus and must be approved indicating acceptability. This is often referred to as communicated risk.

In order to apply a risk management strategy, a procedure or SOP on risk management is typically available within manufacturing companies. This should clearly describe the risk management process and the various risk assessment tools, their application and guidance on how to complete them. The content of any risk management procedure or SOP should align with ISO 14971:2007 Medical Devices - Application of Risk Management to Medical Devices. Controlled templates for PFMEAs etc. also bring consistency and continuity to the process.

Design and Development Planning

It is the manufacturer's responsibility to establish and maintain plans that describe or reference the design and development activities and define responsibilities for implementation. The plans should identify and describe the interaction with different groups or activities that are part of the design and development process. The maintenance of plans to reflect an accurate state as the design and development progresses is also a key factor. The design and development planning is intended to be prospective in nature. It allows risks to be identified earlier and promotes timely delivery of projects.

Design Input Phase

The aims of the Design Input Phase are to (1) define and document the user needs and the intended use of the medical device and (2) translate user needs and the intended use of the packaged device into design input requirements. (E.g. engineering specifications and the product requirements specifications.)

The typical documents required when establishing design inputs include:

- The creation of a formal design description detailing the intended use, user requirements and design inputs. (Note: the design description must align with the design input requirements.)
- A design and development plan which provides an estimation of timelines, resources required, responsibilities, project risks and scope of the project.
- Initial risk assessment which contains the user, design and component risks to be mitigated.

- Design concepts and technology overview.
- Business case report addressing the market size and market opportunity.

FDA 21 CFR Requirements – Design Input

21 CFR Part 820.30(C) Design Input

- *Each manufacturer shall establish and maintain procedures to ensure that the design requirements relating to a device are appropriate and address the intended use of the device, including the needs of the user and patient.*

- *The procedures shall include a mechanism for addressing incomplete, ambiguous, or conflicting requirements.*

- *The design input requirements shall be documented and shall be reviewed and approved by designated individuals.*

Incomplete requirements can have a serious and costly effect on the design and ultimate success of a product. If essential design requirements are omitted in error or otherwise, the impact on quality or functionality may not be detected until validation. This presents an expensive problem that may not be easily rectified. If design requirements are missed, a redesign may be necessary before a design can be released to production, thus causing delays to the project. Furthermore, if modifications are required to tooling, or process equipment, timelines can be impacted greatly. However, the safety and quality of the product must be paramount. Keeping one eye on the user requirements and intended use of the product is an important factor in avoiding gross design requirement failings.

What Is Design Input?

An artist's impression or concept documents do not meet the true intent of design input requirements. The purpose of design input is to create a *complete* set of requirements that are written in a technical manner with an engineering and scientific level of detail. The use of qualitative terms in a concept document is both appropriate and practical. This is often not the case for a document to be used as a basis for design. The language used in the creation of Design inputs also has a profound impact on the direction and scope of a product. If a concept document describes the product to be suitable for "outside use", then there will be requirements with regards to insulation, water ingress and operating temperatures and so on.

Scope

Design input requirements must be comprehensive. This may be quite difficult for manufacturers who are implementing a system of design controls for the first time. Design input requirements fall into three categories with most products having requirements within all three categories including:

(1) Functional requirements detailing the operation of the device.

(2) Performance requirements detailing the performance requirements or expectations of the device in relation to accuracy, speed of response times, battery life, device safety and reliability etc.

(3) Interface requirements specifying features of the device which are critical to compatibility with external systems such as the patient interface.

The scope of design input work depends on the complexity of a device and the risk associated with its use.

Tips for Reviewing Design Input Requirements

The ultimate goal of the design input phase is to gain agreement and approve the requirements formally. At this point, the document is a controlled document and subject to change control. Any updates required at a later date will need to be done through the change control process.

Design Input Requirements Should Be Crystal Clear: For example, a medical device may require use of a built-in battery. It would be important to specify the life expectancy of the battery. To say it has an approximate operating life of 2-3 years is too vague. A better description would be to say it has 2000 hours of operation with a software requirement that logs the number of hours the device is powered on. This mitigates the likelihood of failure during use.

Use of Tolerances: For example, a contact lens may have an outer diameter of 14.00mm. While this is the target/nominal value it cannot be ever accurately achieved. There will always be a degree of variation in the diameter measurement. Applying a tolerance, allows an acceptable range in which the measurement is within specification and accepted. If the diameter is specified as 14.00 ± 0.2mm, designers have a basis for determining how accurate the manufacturing processes have to be. In addition, the specification will allow designers determine if the design meets the intended use.

Industry Standards: Design input requirements should meet or exceed industry standards. Compliance to product specific standards should be considered.

Environment: The operating environment of the device should be specified. Take the example of a cardiac defibrillator. If the device is intended for use on a frontline ambulance it may be used outdoors in cold and damp conditions. On the other hand, use within a hospital setting would require greater control of the temperature range and environmental conditions.

Design Output Phase

The purpose of the design selection(output) phase is to provide a range of design options and solutions with the relevant evidence to show the effectiveness of the same. Often proof of concept (POC) or proof of principle (POP) trials may be used to verify effectiveness of solutions. POC/POP testing can involve making some limited prototypes. Any documents created in the previous phase, design input, should be reviewed and updated if required. There should be no contradictions or gaps between the documented inputs and outputs.

FDA 21 CFR Requirements – Design Output

21 CFR Part 820.30(D) Design Output

- *Each manufacturer shall establish and maintain procedures for defining and documenting design output in terms that allow an adequate evaluation of conformance to design input requirements.*

- *Design output procedures shall contain or make reference to acceptance criteria and shall ensure that those design outputs that are essential for the proper functioning of the device are identified.*

- *Design output shall be documented, reviewed, and approved before release.*

- *The approval, including the date and signature of the individual(s) approving the output, shall be documented.*

During this phase, product specifications (PS) and the device master record (DMR) are generated to define the design output. Planning for process validation and manufacturing begins during this phase often with the creation of a validation master plan (VMP). In any design office or factory setting, a lot of data and paperwork are generated. Therefore, it is important to be able to make the distinction between what is a design output and what is not. The first way of identifying a design output is to verify if it is listed as a task, a deliverable or listed in the design and development plan. If this is the case, then it is classified as a design output. Furthermore, if it describes or defines a design feature, it can also be classed as a design output.

Production Specifications

Production specifications draw upon many documents that are used to manufacture, test, inspect, install, maintain and service a device. They include: (1) component and material specifications, (2) production and process specifications, (3) work instructions and SOPs, (4) quality plans, specifications and procedures, (4) labelling specifications, and (5) packaging specifications.

Design Review

Formal design reviews are critical to the efficacy of design control, and ultimately, the market success of the device. They should be planned for up front in the design development plan. Changes late in the design cycle are much more expensive than those made early on. Design reviews can play an important role in identifying changes in a timely manner and thus prevent costly redesigns close to the launch date. The FDA QSR clearly specifies the need for independent reviewers. Independent reviewers must be far enough removed from the design in order to provide an objective review.

FDA 21 CFR Requirements- Design Review
FDA CFR Part 820.30(E) Design review

- *Each manufacturer shall establish and maintain procedures to ensure that formal documented reviews of the design results are planned and conducted at appropriate phases of the device's design development.*

- *The procedures shall ensure that participants at each design review include representatives of all functions concerned with the design phase being reviewed and an individual(s) who does not have direct responsibility for the design phase being reviewed, as well as any specialists needed.*

- *The results of a design review, including identification of the design, the date, and the individual(s) performing the review, shall be documented in the design history file (the DHF).*

Key goals of design review:

- provide feedback to designers on existing or emerging problems
- assess project progress
- provide confirmation that the project is ready to move on to the next phase of development

Many types of reviews occur during the course of developing a product. Reviews may have both an internal and external focus.

Design Verification, Validation and Transfer Phase

To illustrate the concepts, consider a building design. In a typical scenario, the senior architect establishes the design input requirements and sketches the general appearance and construction of the building, but contractors typically elaborate and interpret the details into practical terms. Verification refers to the checking at each phase to ensure the output meets the design requirements. For example, if a device is designed to take both AC electrical power and a battery (DC power), the design engineer must verify that these are accounted for in the plans and production specifications.

FDA 21 CFR Requirements - Design Verification

FDA CFR Part 820.30(f) Design Verification

- *Each manufacturer shall establish and maintain procedures for verifying the device design.*
- *Design verification shall confirm that the design output meets the design input requirements.*
- *The results of the design verification, including identification of the design, method(s), the date, and the*

individual(s) performing the verification, shall be documented in the Design History File.

The ultimate aim of design verification is to finalise design specification. Examples of verification activities include:

- Design failure modes and effects analysis (DFMEA)
- Fault tree analysis
- Package integrity tests
- Biocompatibility testing
- Bioburden testing of packed products
- Worst case analysis – tolerance stacking of components

Design Validation

Design validation is required for the product to ensure the device meets the user requirements and intended use. Above all, it ensures the device operates reliably and safely. Process validation is required in order to confirm manufacturing specifications and the Device Master Record (DMR).

FDA 21 CFR Requirements - Design Validation FDA CFR 820.30(G) Design Validation

- *Each manufacturer shall establish and maintain procedures for validating the device design.*
- *Design validation shall be performed under defined operating conditions on initial production units, lots, batches, or their equivalents.*
- *Design validation shall ensure that devices conform to defined user needs, intended uses and shall include testing of production units under actual or simulated use conditions.*
- *Design validation shall include software validation and risk analysis, where appropriate.*
- *The results of the design validation, including identification of the design, method(s), the date, and the*

individual(s) performing the validation, shall be documented in the design history file.

Verification examines design outputs at the different phases of the process while design validation confirms that all user needs are achieved even when subject to anticipated sources of variation such as materials, processing equipment, suppliers and so on.

Design Transfer

The purpose of design transfer is to finalise all deliverables for filing with regulatory agencies.

FDA 21 CFR Requirements - Design Transfer
FDA CFR Part 820.30(H) Design Transfer

- *Each manufacturer shall establish and maintain procedures to ensure that the device design is correctly translated into production specifications.*

As the design output is finalised, the design is transferred into production specifications (drawings, manufacturing, test, and inspection procedures). Production specifications must ensure that manufactured devices are consistently and reliably produced within product and process capabilities, meeting all quality requirements.

Design Control Deliverables

This section provides a non-exhaustive list of design documentation deliverables. A brief description of each is provided. This list can be used as a checklist for the design control process or as supplementary information of key activities outlined previously.

Validation Master Plan (VMP): A validation master plan should be written as soon as the project begins. It should describe the product to be manufactured and the process technology. A VMP will also contain generic material such as an outline of the validation approach and the types of validation e.g. prospective, concurrent and so on.

External Requirements: External requirements refer to regulations and industry standards that are relevant to a new product. At the design input phase a list of documents should be created in order to capture essential requirements as early as possible.

Design Development Plan: A design and development plan is an overarching document that describes the design and development, responsibilities, timelines and project scope, list and schedule of major tasks and the phase review details such as the timing and approval requirements.

Product Specification: The product specification is a design output document that is built over the course of the project. Not all information will be final in the early phases, however, having an early draft will help focus minds and generate the right activity in order to define target dimensions, physical attributes and tolerances.

Stability Testing: A document containing a summary of results, testing and analysis should be created and filed as part of the DHF.

Device Master Record: A DMR is an output document and should be available at the design transfer phase. It is a comprehensive list referencing all work instructions, test procedures, test specifications, manufacturing specifications and finished product specifications required to manufacture the product.

Test Method Validations: A list of all validated test methods (functional, analytical, physical etc.) should be available to file in the DHF.

Design History File: The DHF is a repository for all of the documentation generated as a result of the design control process. The DHF serves as a complete record of the design.

Design Control Process via Web-Based Systems

In recent years some companies have entered the market offering web-based design control processes. As mentioned earlier, there are a large amount of documents created during the design control process. Most of the documentation generated is subject to change control and therefore requires review and approval. As with traditional hardcopy approval, this can be time-consuming and complex if approvers are based across different departments or drawn internationally.

All documents also form part of the design history file. Therefore, the proper filing and availability of documents is an important source of concern. The use of an electronic system may mitigate some of these concerns.

Furthermore, some web-based solutions offer integration with existing electronic documentation systems or integration with quality system software such as CAPA software and/or deviation management software.

11

RISK MANAGEMENT

This chapter covers the risk management requirements of ISO 13485 and the essential requirements of the 93/42/EEC Medical Device Directive, the 98/79/EC IVD Directive and 90/385/ECC – The Active Implantable Medical Devices Directive (AIMDI).

Typically, compliance to the above requirements is met by meeting ISO 14971:2009. This chapter provides an overview of risk management activities for regulated medical devices. It may also be used as a framework for performing risk management for non-regulated products.

Definitions

Risk Management: Systematic application of

management policies, procedures and practices to the tasks

of analysing, evaluating, controlling and monitoring risk.

Hazard: A hazard is a potential source of harm (physical injury or damage to the health of people, damage to property or the environment).

Failure Modes: Product or process failures that may lead to a hazard.

Risk: The combination of the probability that harm will

occur and the severity of that harm.

Residual Risk: Risk remaining after risk control measures have been taken.

Risk Analysis: The use of available information to identify hazards and estimate risk.

Risk Reduction: Processes, controls or information that will reduce risk.

Risk Evaluation: The process of comparing the estimated risk against given risk criteria to determine the acceptability of the risk.

Risk Control: The process in which decisions are made and measures implemented by which risks are reduced to, or maintained within, specified levels.

Severity: A measure of the possible consequences of a hazard.

Probability of Failure: An estimate of the likelihood of failure.

Detection of Failure: An estimate of the likelihood of detection.

Risk Priority Number (RPN): The product of ratings on occurrence, severity and detection.

CAPA: Corrective and Preventative Action.

ALARP: As Low As is Reasonably Practicable.

Verification Plan: A confirmation through the provision of objective evidence, that the specified requirements have been fulfilled.

Failure Modes and Effects Analysis, FMEA: A formal risk assessment methodology for identifying potential failure
modes, and assigning numerical values to the severity, likelihood of occurrence, and likelihood of escaped detection to failure modes in order to quantify risk.

Risk Analysis

Risk analysis can be performed using a variety of methodologies such as FMEA/FMECA, HAZOP,

HACCP, or other methods appropriate to the design and function of the product. A common methodology for risk analysis for regulated products is FMEA. The analysis can be grouped into a product category of similar established device technology. In estimating risk(s) for each hazard, information may be obtained from the following sources:

- Published standards
- Clinical trial data
- Technical data
- Field data from similar products
- Usability tests
- Results of investigations, e.g. CA/PA investigations, etc.
- Expert opinion
- External assessments or audits

Failure Mode Effects and Analysis (FMEA)

FMEA consists of listing all the potential failure modes associated with the processes followed by a corresponding list of all the possible causes and effects of each potential failure mode. The impact on the patient, operator, environment, process, handlers and business should be considered.

Severity (S): the severity score addresses how severe the effect of this failure is on all/or one of the following: the patient, operator, environment, process, handlers and business.

Occurrence (O): refers to how often the failure is expected to occur. This can be rated either subjectively (frequent, rare etc.) or via the number of units affected depending on the risk being assessed.

Detection (D): can the problem be detected by the user or patient before it does any damage?

Risk Control

Where risks are identified as unacceptable, risk control measures must be determined to reduce the risk prior to the process or system being implemented. A number of actions can be taken in order to further reduce risk including: (1) changing the design to reduce risk, introducing protective measures in the device or the manufacturing process, (3) inserting a warning statement into the instructions for use (IFU).

Risks scored as "investigate further risk reduction" should be examined to determine whether it is practicable to reduce the risk further. The risk should be reduced to as low as is reasonably practicable, (aka ALARP) taking into account the benefits of accepting the risk and the practicability of implementation. If risks classed as "investigate further risk reduction" are already at ALARP, no further risk reduction is necessary.

Residual Risk Evaluation

After risk control measures are applied, a new risk assessment will be carried out to determine residual risks. If the residual risk is not judged acceptable then further risk control measures will be applied.

If the residual risk is not judged acceptable and further risk control is not practicable then the team may perform a risk/benefit analysis by evaluating data and literature on the medical benefits of the intended use to determine if they outweigh the risk. If this evidence does not support the conclusion that the medical benefits outweigh the residual risk, then the risk remains unacceptable. This analysis should be recorded and approved by both the risk management team and senior site management.

EN ISO 14971: 2009– Characteristics - Annex C

Annex C contains several questions that can be used to identify medical device characteristics that could impact upon safety:

What is the intended use and how is the medical device to be used?

Is the medical device intended to be implanted?

Is the medical device intended to be in contact with the patient or other persons?

What materials or components are utilised in the medical device or are used with, or are in contact with, the medical device?

USEFUL REFERENCES

EN ISO 13485:2012 Medical Devices-Quality
Management Systems-Requirements for Regulatory
Purposes

ISO 9000:2005 Quality Management Systems

ISO/TR 14969:2004, Quality Management Systems-
Guidance on the Application of ISO 13485

ISO 14971:2007- Risk Management

ISO 14001:2004 Environmental Management Systems

ISO 19011:2011 Auditing Management Systems

OHSAS 18001 Occupational Health and Safety
Management

ISO 15223-1:2012 Symbols to Be Used with Medical
Device Labels, Labelling and Information to Be Supplied

ISO 18113-1:2009 In Vitro Diagnostics for Medical
Devices

COMMON ACRONYMS

CFR: Code of Federal Regulation

EU: European Union

EC: European Council

EMEA: European Medicines Agency

ISO: International Standards Organisation

MDD: Medical Device Directive

QMS: Quality Management System

FDA: Food and Drug Administration (US regulatory body)

HPRA: Health Products Regulatory Authority

cGMP: Current Good Manufacturing Practice

www.ingramcontent.com/pod-product-compliance
Lightning Source LLC
Chambersburg PA
CBHW070029210526
45170CB00012B/508